J. WESTON
WALCH
PUBLISHER

GRAPHIC ORGANIZERS
FOR GEOMETRY

Daniel J. Barnekow
illustrated by Roman Laszok

User's Guide
to
Walch Reproducible Books

Purchasers of this book are granted the right to reproduce all pages where this symbol appears.

This permission is limited to a single teacher, for classroom use only.

Any questions regarding this policy or requests to purchase further reproduction rights should be addressed to:

Permissions Editor
J. Weston Walch, Publisher
321 Valley Street • P. O. Box 658
Portland, Maine 04104-0658

1   2   3   4   5   6   7   8   9   10
ISBN 0-8251-4344-6
Copyright © 2002
J. Weston Walch, Publisher
P. O. Box 658 • Portland, Maine 04104-0658
www.walch.com

Printed in the United States of America

Contents

Preface . . . . *v*

To the Teacher . . . . *vii*

I. Understanding Geometry 1
Teaching Notes 2
Student Activity Sheets 4
    UG-1: What Is Geometry? 4
    UG-2: Why Geometry Is Important 5
    UG-3: Geometry in Everyday Life 6
    UG-4: A Famous Geometer 7
    UG-5: Terms Used in Geometry 8

II. Two-Dimensional Geometric Shapes 9
Teaching Notes 10
Student Activity Sheets 13
    TGS-1: What Is a Flat Shape? 13
    TGS-2: Characteristics of Two-Dimensional Shapes 14
    TGS-3: Circles 15
    TGS-4: Polygons 16
    TGS-5: Quadrilaterals and Parallelograms 17
    TGS-6: Triangles 18
    TGS-7: The Pythagorean Theorem 19
    TGS-8: Perimeter 20
    TGS-9: Area 21

III. Three-Dimensional Geometric Shapes 23
Teaching Notes 24
Student Activity Sheets 28
    THGS-1: What Is a Solid? 28
    THGS-2: Characteristics of Three-Dimensional Shapes 29
    THGS-3: Polyhedrons 30
    THGS-4: Cubes 31
    THGS-5: Prisms 32
    THGS-6: Pyramids 33
    THGS-7: Spheres 34
    THGS-8: Cylinders 35
    THGS-9: Cones 36
    THGS-10: Surface Area 37
    THGS-11: Volume 38

IV. Coordinate Geometry 39
    Teaching Notes 40
    Student Activity Sheets 41
        CG-1: What Is Coordinate Geometry? 41
        CG-2: The Coordinate Plane 42

V. Transformations 43
    Teaching Notes 44
    Student Activity Sheets 45
        T-1: Symmetry 45
        T-2: Similarity 46
        T-3: Congruence 47
        T-4: Rotations, Slides, and Flips 48

VI. Communication and Problem Solving 49
    Teaching Notes 50
    Student Activity Sheets 52
        CPS-1: Problem-Solving Strategies 52
        CPS-2: K-W-L Chart 53
        CPS-3: Problem-Solving Myths 54
        CPS-4: Myself as a Problem-Solver 55
        CPS-5: Geometric Communication Skills 56
        CPS-6: A Letter Explaining a Geometric Concept 57
        CPS-7: Working in a Group 58

VII. Connections 59
    Teaching Notes 60
    Student Activity Sheets 61
        C-1: The Branches of Geometry 61
        C-2: A Geometry Problems Constellation 62
        C-3: How I Use Geometry in Other Subjects 63
        C-4: The Golden Rectangle in Art and Architecture 64

VIII. Additional Graphic Tools 65
    Teaching Notes 66
    Student Activity Sheets 67
        AGT-1: Geometry Symbol Log 67
        AGT-2: Geometry Formula Log 68
        AGT-3: Personal Geometry Glossary 69
        AGT-4: Geometry Use Log 70
        AGT-5: Coordinate Plane Template 71
        AGT-6: The Six Questions 72
        AGT-7: Venn Diagram Template 73
        AGT-8: 10×10 Grid 74
        AGT-9: 100×100 Grid 75
        AGT-10: Dealing with Math Anxiety 76

Preface

This book was conceived as a kind of "Swiss Army knife" for middle school geometry teachers. The idea was to create a teaching resource that would be there when you need it, help you accomplish a wide variety of tasks, and get you out of some tough spots.

That was the idea. The result is this book. Between its covers, you'll find a quick, prepared, practical assignment for virtually every major topic you're likely to cover in your geometry classes. What's more, each assignment is designed to teach, reinforce, and extend the key ideas of the geometry curriculum. This was accomplished by combining a teaching technique that has proven pedagogical value with national mathematics curriculum standards.

You can draw on the student activity sheets in *Graphic Organizers for Geometry* nearly any school day. At your fingertips you'll have a ready-made, easy-to-use, pedagogically valuable lesson, on nearly any topic. The lessons will be popular with students and are suitable for in-class or homework assignments.

We wish you, and your students, the best of luck in your endeavors.

My appreciation is extended to the geometry teachers and, especially, to their kids, without whose insight this book would not be possible.

—Daniel J. Barnekow

To the Teacher

A teacher once described graphic organizers as "sophisticated doodles." In a way, he was right. In fact, that may be the best way for you to think about graphic organizers and to present them to your students. You can find many jargon-laden articles and books that analyze graphic organizers, put forth new taxonomies, and labor to link them to psychological dynamics. These have their place, of course, but graphic organizers—essentially a simple teaching tool—have been overanalyzed, with the net effect of confusing rather than enlightening educators.

Graphic Organizers for Geometry is designed to cut through the jargon and give you a practical tool that you can put to use immediately. Spend a little time reading this introduction and thumbing through the graphic organizers, and you'll be ready to go.

Understanding Graphic Organizers

On a practical, classroom level, all you need to know about graphic organizers can be summed up in a few key points. As you use this book, or use graphic organizers in any educational context, keep these ideas in mind:

Graphic organizers are simply ways to organize information visually. This is a simple, straightforward, and accurate description of graphic organizers.

Graphic organizers are nearly always appropriate. Many people tend to think in visual terms, so graphic organizers are an appropriate way to organize information.

Graphic organizers come in many forms. Many attempts have been made to categorize graphic organizers and to identify them by type. You've probably heard of sequence chains, concept maps, webs, flowcharts, Venn diagrams, and so on. (You'll find many of these in this book.) But some of the best graphic organizers are combinations of these standard forms, and some are utterly unique.

Graphic organizers are never right or wrong, only better or worse. As long as the facts presented, and their interrelationships, are correct, there are no "wrong" graphic organizers. However, some do a better job of presenting the same information than others.

Graphic organizers are not communicative, but conceptual. They are tools that help students acquire knowledge, not impart it. Obviously, graphic organizers are excellent communication tools, but in the classroom, you should focus on using them as a way for students to learn, and not as a way for students to show you what they have learned.

Graphic organizers are concept-driven. The form of a graphic organizer should follow its function, not vice versa.

Content and Organization: Major Fields, Key Concepts, and Main Ideas

This book covers a wide range of middle school geometry topics, as a glance at the table of contents will show. The major sections of the book correspond to *major fields* taught in geometry classes at the middle level. Within each major field, the graphic organizers emphasize *key concepts*. Each graphic organizer focuses on the *main ideas* of each key concept.

This organization enables you to use these graphic organizers throughout the year to help students achieve the principal learning objectives of your middle-level geometry class.

Correlation to National Standards

The content and organization of *Graphic Organizers for Geometry* was inspired and guided by a respected set of national standards: *Principles and Standards for School Mathematics*, issued by the National Council of Teachers of Mathematics. The major sections of the book correspond to these major content standards. Each graphic organizer in this book supports one or more specific NCTM standards for this level.

How to Use the Graphic Organizers in This Book

Of course, you can use these graphic organizers any way you see fit—for basal instruction, review, and extension and enrichment. You can have students work in pairs or small groups to complete them. They function equally well as homework and in-class assignments, and are also excellent guides for classroom discussion.

A Lesson Cycle for Individual Graphic Organizers: Educators have learned that following a few simple steps will help their students get the most out of their graphic organizers. These steps, tailored to the content of this book, are presented below in—well, a graphic organizer!

---

1. Familiarize yourself with the graphic organizer and the teaching notes for it.

↓

2. Explain, or remind students of, what graphic organizers are and why they're worthwhile. Emphasize the importance of organizing information.

↓

3. Present the specific graphic organizer. Point out its subject, its organizational framework, and the introduction, direction line, and questions.

↓

4. Model using the graphic organizer. Use examples. Consider giving students an example of what to include in each cell. If the graphic organizer calls for students to choose the topic, provide them with options.

↓

5. Assign the graphic organizer as an individual, paired, or group activity.

↓

6. Review students' work. Use the Key Questions in the Teaching Notes to generate classroom discussion or extend individual student learning.

How to Use the Graphic Organizers as a Set

Because *Graphic Organizers for Geometry* has fairly comprehensive topic coverage, inspired by national standards, you can use the organizers as curricular signposts, correlating them to the main points in your curriculum.

Also, because these graphic organizers emphasize main ideas in the key concepts of the major fields of student study, they are excellent candidates for inclusion in student portfolios.

Teaching Notes

Teaching notes for each graphic organizer in this book are provided at the beginning of each section. The notes are organized in the following format:

[number]
[Title of Graphic Organizer]

<u>Objective</u>
   Identifies the major learning objective of the graphic organizer.

<u>Key Questions</u>
   Key questions that generate classroom discussion, guide students in achieving the learning objective, and extend teaching about the subject of the graphic organizer.

<u>Usage Notes</u>
   Tips and techniques for using the graphic organizer in the most effective manner.

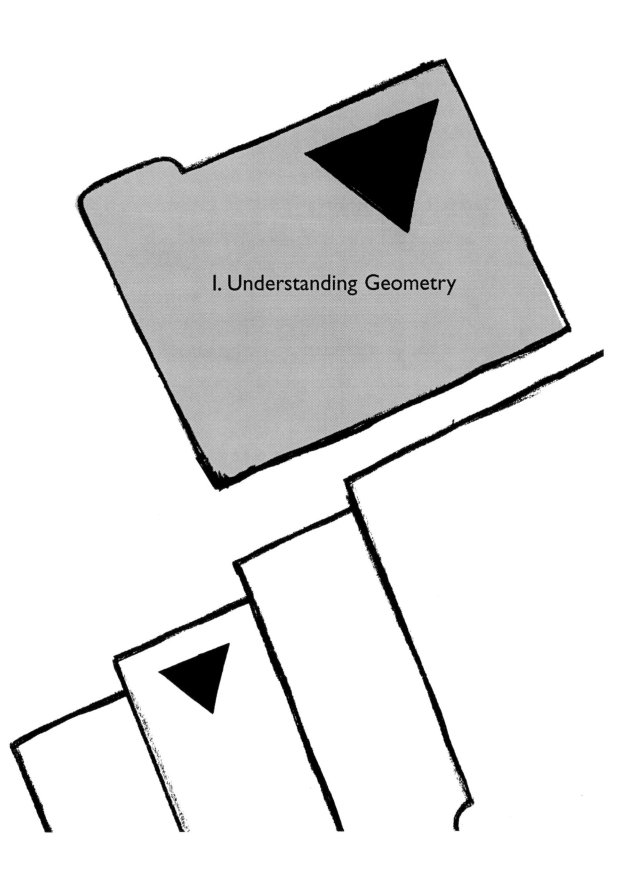

I. Understanding Geometry

UG-1: What Is Geometry?

Objective: Students will answer 12 fundamental questions about geometry, then ask and answer 6 more.

Key Questions:

- Questions on the graphic organizer.

Usage Notes: Have students ask the questions they generate aloud, then use them as the basis for class discussions.

UG-2: Why Geometry Is Important

Objective: Students will identify the importance of geometry in several areas.

Key Questions:

- How is geometry used in each area?
- Who uses geometry in each area?

Usage Notes:

- To extend, have students interview appropriate adults in various fields about their use of geometry and report what they learn to the class.
- Have students update this graphic organizer throughout the year as they learn new uses for geometry.

UG-3: Geometry in Everyday Life

Objective: Students will identify and describe the use of geometry in their community.

Key Questions:

- What items did you see?
- How did geometry play a part in each one?
- Why is it important to learn about geometry?

Usage Notes: Emphasize the fundamental role geometry plays in the human-created world.

UG-4: A Famous Geometer

Objective: Students will investigate and report on the life of a famous geometer.

Key Questions:

- What contributions did your geometer make?
- Why are these contributions important?
- Would you like to work as a geometer? Why or why not?

Usage Notes: You might want to suggest the following as topics: Archimedes, Thales, Pythagoras, Euclid, René Descartes, Pierre de Fermat, Carl Friedrich Gauss, Janos Bolyai, Nikolai Lobachevsky, Georg Friedrich Bernhard Riemann.

UG-5: Terms Used in Geometry

Objective: Students will define and provide examples or illustrations of terms used in geometry.

Key Questions:

- What does each term mean?
- What is an example that shows each term in action?
- How does the etymology of each term help you remember its meaning?

Usage Notes:

- Have students keep this graphic organizer in their notebooks as a quick-reference tool.
- Work with the class as a whole so that every student hears and takes notes on the etymology of each term.

Name_____ Date _____

UG-1: What Is Geometry?

Introduction: Geometry is one of the chief branches of mathematics. It is concerned with the shape, size, and position of geometric figures, both two- and three-dimensional. You can quickly learn more about geometry by asking—and answering—a few questions.

Directions: Conduct research to answer questions 1 and 2 in each shape.

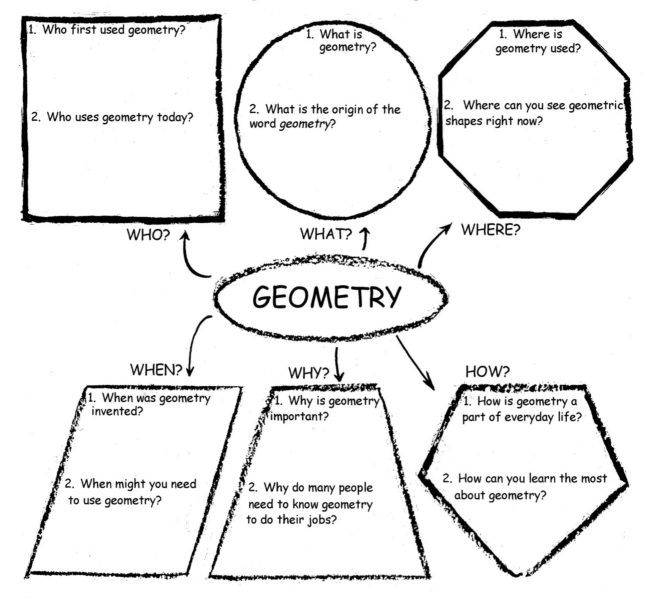

Taking Another Step: Ask six questions of your own about geometry and write them on the back of this sheet. Then trade your sheet with a partner and answer each other's questions.

4   *Graphic Organizers for Geometry*

Name_____ Date _____

UG-2: Why Geometry Is Important

Introduction: By now, you have probably been told by a teacher or two that geometry is important. Anyone who says that is absolutely right. But are you sure you understand *why* geometry is so important? To make sure, complete the diagram below.

Directions: Complete the organizer by writing at least three ways geometry is important in each category.

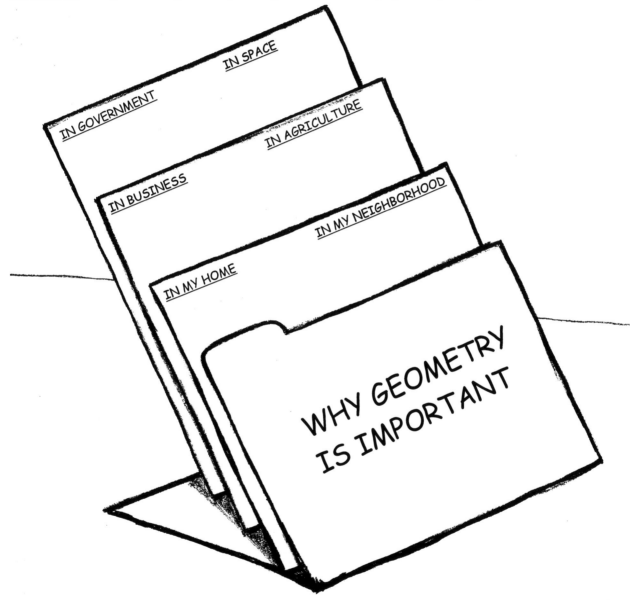

Taking Another Step: Choose one of the categories listed in the diagram, or make up one of your own. Then write a paragraph that explains why geometry is so important in that category.

Name _____ Date _____ | **Student Activity Sheet**

UG-3: Geometry in Everyday Life

Introduction: Many students feel that math has little to do with real life. In fact, most of the real lives that you and your classmates live are based, to a greater or lesser degree, on math. This is especially easy to see when you think about geometry. Your world is filled with lines and shapes put together in a way that gives the world an order. Your home, your clothes, the road you take to school, your school, and the many things you use every day are based on geometry.

Directions: Spend a day noticing how geometry is used in the things that you use and see every day. For example, you might notice straight lines, right angles, and circles on a sheet of notebook paper. You might see the triangular supports of a bridge in your community. As you notice the role of geometry in your life, fill in the log below.

Geometry in My World		
Item	**How Geometry Plays a Part**	**Notes**
1.		
2.		
3.		
4.		
5.		
6.		
7.		
8.		
9.		
10.		

Taking Another Step: Take a specific concept from geometry. On the back of this sheet, write a paragraph that explains how it is used in the real world.

Name _____ Date _____ | **Student Activity Sheet**

UG-4: A Famous Geometer

Introduction: Are you a geometer? If not, you soon will be. A geometer is "someone skilled in geometry." Another name for geometer is "geometrician." You can be a geometer—or a geometrician, if you prefer—with just a little effort. A good way to start is to learn about a famous geometer. And, who knows? Someday a student might be learning about *you!*

Directions: Conduct research to learn about a famous geometer. Use what you learn to complete the table below.

The Life and Work of a Famous Geometer	
Name	
Life dates	
Where geometer lived and worked (country, city, etc.)	
Major works	
Important contributions to geometry	
Interesting personal information	
Additional information	

Taking Another Step: On the back of this sheet, create a piece of artwork to illustrate your table. You might sketch a picture of the geometer, draw a geometric figure to illustrate his or her work, make a rough map to show where your geometer lived or lives, or do anything else that is appropriate.

Name _____ Date _____   **Student Activity Sheet**

UG-5: Terms Used in Geometry

Introduction: Hexagon, solid, diameter, surface area, perpendicular . . . geometry has its own vocabulary. To understand geometry, you need to understand the terminology of geometry. Most of the words aren't that difficult. In fact, you probably already know a lot of them. (You already know better than to put a *square* peg in a *round* hole, don't you?!) To help you understand some new terms, follow the directions.

Directions: Fill in the "book" as you learn new geometric terms. In the Notes column, write down something that will help you remember what each term means.

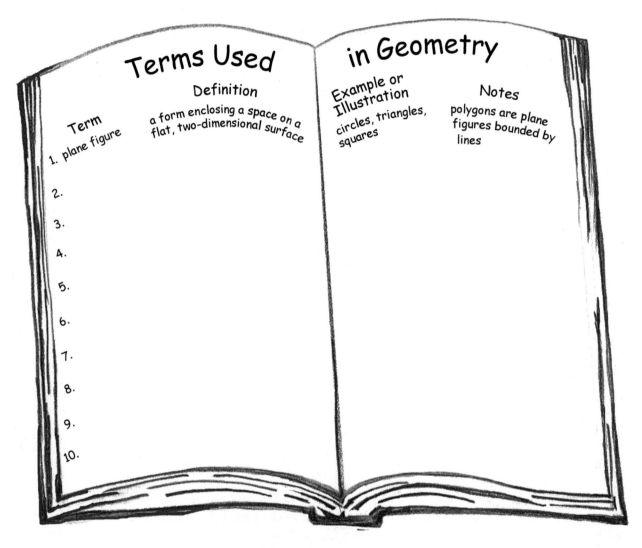

Terms Used in Geometry

Term
1. plane figure
2.
3.
4.
5.
6.
7.
8.
9.
10.

Definition
a form enclosing a space on a flat, two-dimensional surface

Example or Illustration
circles, triangles, squares

Notes
polygons are plane figures bounded by lines

Taking Another Step: Use a dictionary to find the etymology, or origin, of three of the terms above. On the back of this sheet, explain why each of these terms is named as it is.

8   *Graphic Organizers for Geometry*

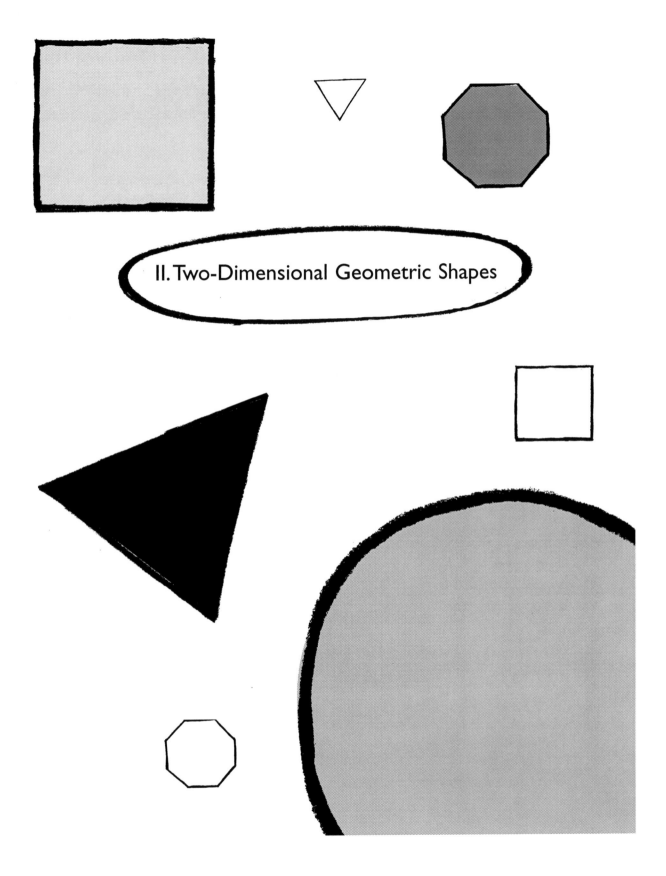

II. Two-Dimensional Geometric Shapes

TGS-1: What Is a Flat Shape?

Objective: Students will define the term *two-dimensional shape* and identify the essential characteristics of two-dimensional shapes.

Key Questions:

- Question forms of the headings in graphic organizer.
- What are the characteristics of each shape?

Usage Notes: Explain to students that two-dimensional shapes or flat shapes are also known as *plane figures*.

TGS-2: Characteristics of Two-Dimensional Shapes

Objective: Students will identify a variety of two-dimensional shapes, identify their characteristics, and sketch them.

Key Questions:

- What shapes did you identify?
- What are the characteristics of each shape?

Usage Notes: Emphasize the "characteristics" cell for each shape; students should refer to qualities of angles, diagonals, sides, etc.

TGS-3: Circles

Objective: Students will define the term *circle* and analyze the key characteristics of circles.

Key Questions:

- What is a circle?
- What is diameter? radius?
- What is circumference?
- What is an arc? a chord?
- What is a secant? a tangent?

Usage Notes:

- Encourage students to record additional key facts about circles on the diagram.
- Offer a mini-lecture about pi (π) before students complete the Taking Another Step activity.

TGS-4: Polygons

Objective: Students will define the term *polygon* and identify and discuss the various types of polygons and the terms associated with them.

Key Questions:
- What is a polygon?
- What defines the different types of polygons?
- What are the sides, perimeter, vertices, and angles of a polygon?

Usage Notes: Have students sketch examples of the different polygons identified in the graphic organizer.

TGS-5: Quadrilaterals and Parallelograms

Objective: Students will define the terms *quadrilateral* and *parallelogram* and compare and contrast fundamental types of parallelograms.

Key Questions:
- What is the difference between a parallelogram and a quadrilateral?
- What is a rectangle? a square? a rhombus?
- Are all rectangles squares? Are all squares rectangles? Are all rhombuses parallelograms?

Usage Notes: Have students sketch examples of the different parallelograms identified in the graphic organizer.

TGS-6: Triangles

Objective: Students will define the term *triangle* and identify key information about triangles.

Key Questions:
- Question forms of the headings in graphic organizer

Usage Notes: Have students sketch examples of the different triangles identified in the graphic organizer.

TGS-7: The Pythagorean Theorem

Objective: Students will identify and describe the Pythagorean theorem.

Key Questions:
- What is the Pythagorean theorem?
- To what does it apply?
- What formula expresses it?
- Why might it be useful?

Usage Notes: For the Taking Another Step activity, explain to students that Egyptians used ropes stretched into triangles to form right angles to lay out their fields. The relationship of the lengths of the rope sides of the triangle ensured the creation of a right angle.

TGS-8: Perimeter

Objective: Students will identify perimeter and explain how to determine the perimeter of various plane figures.

Key Questions:
- What is perimeter?
- How do you determine the perimeter of each of the figures?
- Why is perimeter so called?

Usage Notes: Have students share their results from the Taking Another Step activity. Emphasize that there are many real-world uses for perimeter.

TGS-9: Area

Objective: Students will identify area and explain how to determine the area of various plane figures.

Key Questions:
- What is area?
- How do you determine the area of each of the figures?

Usage Notes: Have students share their results from the Taking Another Step activity. Emphasize that there are many real-world uses for area.

Name_____ Date _____

TGS-1: What Is a Flat Shape?

Introduction: If you think about it, you'll realize that you started learning geometry a long time ago. You learned the word "shape" when you were a little kid. You learned the names of common shapes—square, triangle, circle—when you weren't much older. Now, you need to learn a little more about shapes.

Start by learning that in geometry there are two basic kinds of shapes. *Two-dimensional shapes,* or *flat shapes,* make up one of these two types. Learn more about them by following the directions below.

Directions: Use your textbook, talk to your teacher and classmates, and conduct other research to complete the diagram.

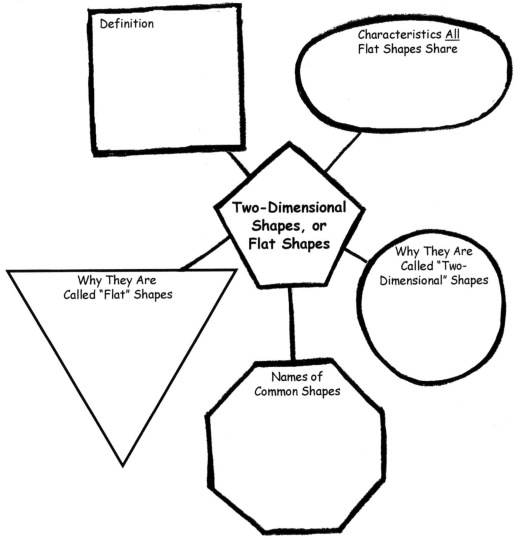

Definition

Characteristics <u>All</u> Flat Shapes Share

Two-Dimensional Shapes, or Flat Shapes

Why They Are Called "Two-Dimensional" Shapes

Why They Are Called "Flat" Shapes

Names of Common Shapes

Taking Another Step: You know that one of the two basic types of shapes in geometry is the two-dimensional, or flat, shape. What do you think the other type is? Write your answer on the back of this sheet. Do your best to write a definition of the type of shape you come up with.

                   Graphic Organizers for Geometry

Name_____ Date _____ | **Student Activity Sheet**

TGS-2: Characteristics of Two-Dimensional Shapes

Introduction: One of the basic types of shapes in geometry is the **two-dimensional shape,** or **flat shape.** There are many different types of flat shapes.

Directions: Complete and expand the diagram by adding boxes to the vertical line as you learn about different types of two-dimensional shapes.

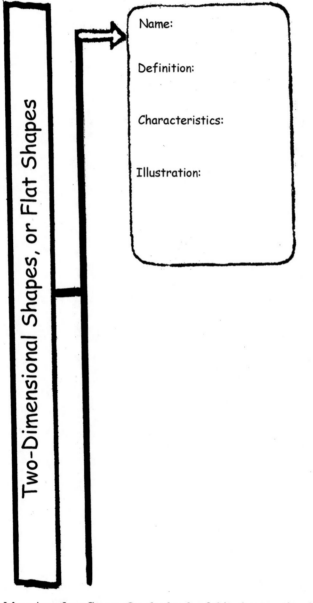

Name:

Definition:

Characteristics:

Illustration:

Two-Dimensional Shapes, or Flat Shapes

Taking Another Step: On the back of this sheet, write down the names of as many different two-dimensional shapes as you can find in your classroom or in your home.

Name_____ Date _____ |

TGS-3: Circles

Introduction: **Circles** are common shapes. Actually, they're *very* common shapes. They appear in nature and in things people create. They are also useful shapes—*very* useful shapes. Because they are very common, and very useful, it pays to know all you can about circles. Knowing about circles well help you be a "well-rounded" person!

Directions: Use your textbook, talk to your teacher and classmates, and conduct other research to complete the diagram.

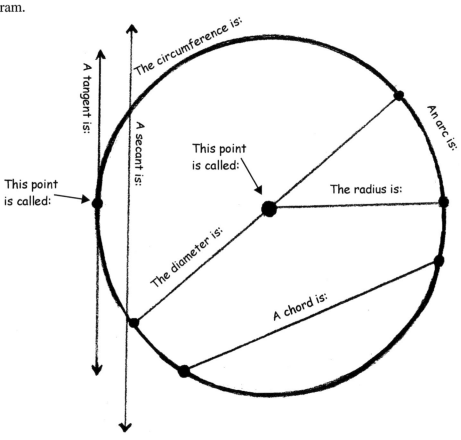

Taking Another Step: Complete the table.

	Pi
Symbol	
Definition	
Value	
Why pi is useful	

       Graphic Organizers for Geometry

Name_____ Date _____ | **Student Activity Sheet**

TGS-4: Polygons

Introduction: Do you know what a **polygon** is? Here's a hint: *poly* means "many." So a polygon has many somethings. But many what? To find this out, follow the directions below.

Directions: Complete the diagram by writing definitions in the boxes.

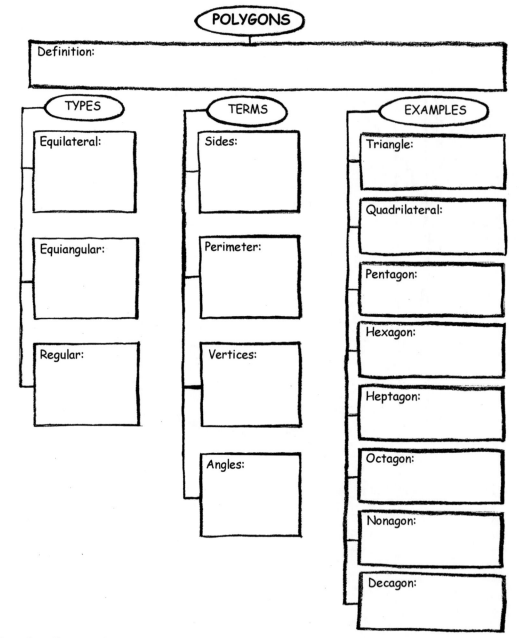

Taking Another Step: Why are polygons called polygons? Write your answer on the back of this sheet. (Remember, *poly* means "many.")

                   Graphic Organizers for Geometry

Name_____ Date _____ | **Student Activity Sheet**

TGS-5: Quadrilaterals and Parallelograms

Introduction: In Greek, *quad* means "four" and *lateral* means "side." So what do you think a **quadrilateral** is? That's right: something with four sides. To be more exact:

> A **quadrilateral** is a polygon with four sides. In other words, it's a flat shape with four straight sides.

One cool thing about quadrilaterals is that they come in an infinite number of shapes and sizes. (Spend just a minute drawing four-sided polygons, and see how many different shapes you can come up with.)

There is one kind of quadrilateral that is of special interest: the **parallelogram.**

Directions: The diagram below shows the relationships among types of parallelograms. Complete it by writing a definition of parallelograms, and by writing definitions and descriptions of the other three terms.

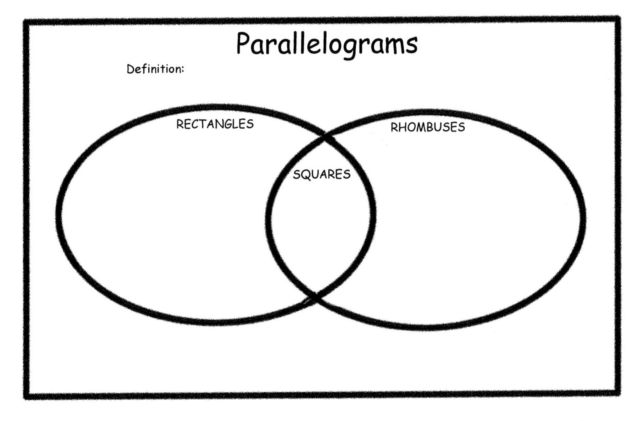

Taking Another Step: Use the four terms below to write four true statements about parallelograms. Your statements must take these forms: "All *x* are *y*" and "Some *x* are *y*, but not all *y* are *x*."

parallelograms   rectangles   squares   rhombuses

          Graphic Organizers for Geometry

Name_____ Date _____ | **Student Activity Sheet**

TGS-6: Triangles

Introduction: **Triangles** are **polygons** with three sides. They are the simplest of polygons. (You couldn't make a polygon with just two sides, right?) Triangles are simple, but they are also very useful. To understand geometry, you need to learn as much as you can about triangles.

Directions: Use your textbook, talk to your teacher and classmates, and conduct other research to complete the diagram.

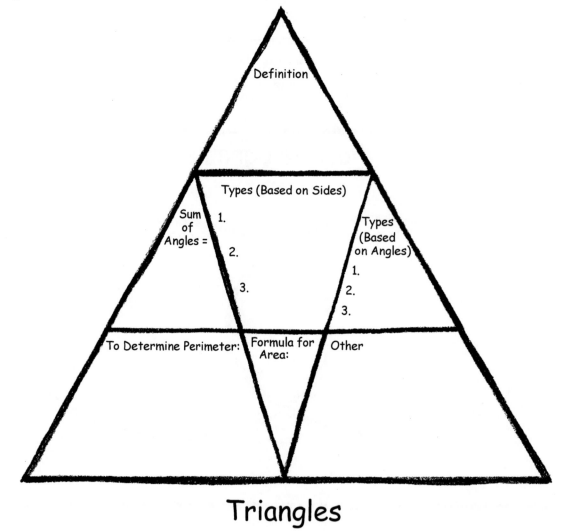

Taking Another Step: How many different triangles can you find in the diagram?

Name_____ Date _____ | **Student Activity Sheet**

TGS-7: The Pythagorean Theorem

Introduction: The **Pythagorean theorem** is named for Pythagoras, a Greek philosopher and mathematician who lived about 2,600 years ago. Pythagoras put together a theorem about the relationships in a right triangle. A **theorem** is a statement that has been proved true. Although this theorem is named for Pythagoras, the ideas it expresses were known by the ancient Egyptians and Chinese.

Directions: As you learn about the Pythagorean theorem, complete the diagram.

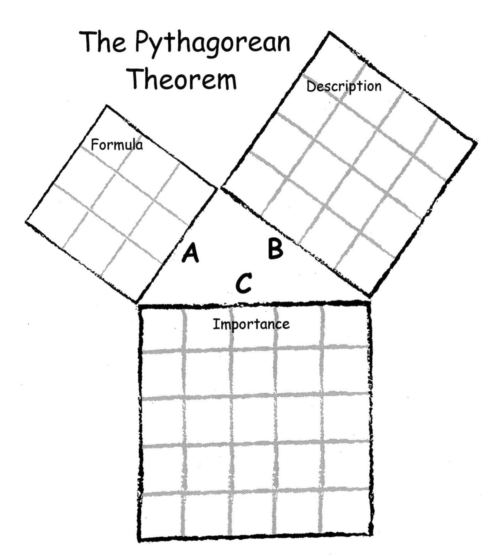

Taking Another Step: The ideas expressed in the Pythagorean theorem were understood by the ancient Egyptians. They used these ideas to help them lay out farm fields. On the back of this sheet, write a paragraph explaining how you think the ideas behind the Pythagorean theorem would be useful for this purpose.

19   *Graphic Organizers for Geometry*

Name_____ Date _____

TGS-8: Perimeter

Introduction: Our word **perimeter** comes from two Greek words that mean "measure" and "around." If you already know what perimeter means, you can see how the name fits. If you're just learning about perimeter, remembering what the name means will help you—a lot!

Directions: Complete the table.

Perimeter	
Definition	
Determining Perimeter	
Figure	**Formula or Process for Determining Perimeter**
circle	
polygon	
parallelogram	
rectangle	
square	
rhombus	
triangle	
other	
other	
other	
other	

Taking Another Step: Think of three real-life situations in which knowing the perimeter of something is important. On the back of this sheet, tell about each one.

 Graphic Organizers for Geometry

Name _____ Date _____

TGS-9: Area

Introduction: A key thing to remember about **area** is that it is expressed in **square measure.** That is, the area of a flat shape is never, say, 25 inches, or feet, or miles. It's 25 *square* inches, or *square* feet, or *square* miles. Remember this:

> The unit for area is the square.

Directions: Complete the table.

Area	
Definition	
Determining Area	
Figure	**Formula or Process for Determining Area**
circle	
polygon	
parallelogram	
rectangle	
square	
rhombus	
triangle	
other	
other	
other	
other	

Taking Another Step: Think of three real-life situations in which knowing the area of something is important. On the back of this sheet, tell about each one.

   Graphic Organizers for Geometry

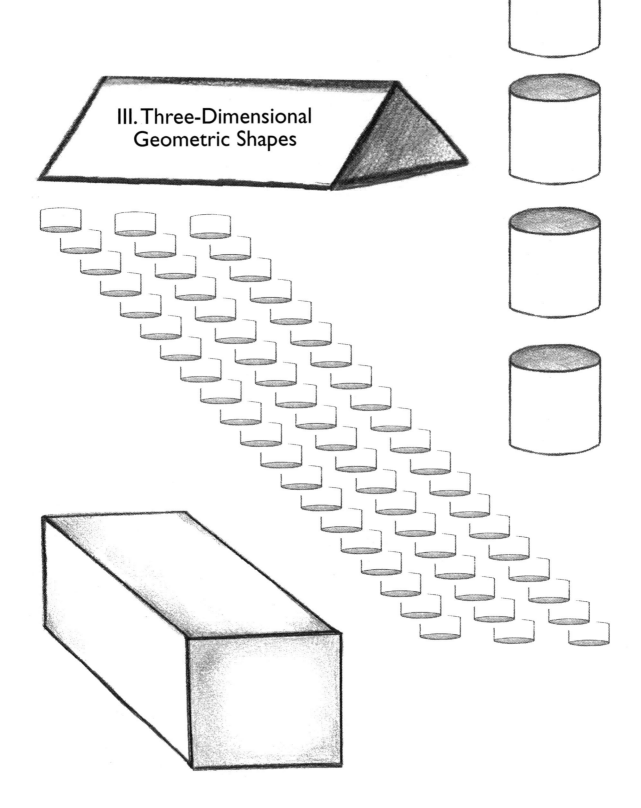

III. Three-Dimensional
Geometric Shapes

THGS-1: What Is a Solid?

Objective: Students will define the term *solid,* identify common solids, and identify their fundamental characteristics.

Key Questions:
- What is a solid?
- What characteristics do all solids share?
- How are they different from flat shapes?

Usage Notes: Explain to students that this is fundamental information that they should work to memorize.

THGS-2: Characteristics of Three-Dimensional Shapes

Objective: Students will identify a variety of three-dimensional shapes, identify their characteristics, and sketch them.

Key Questions:
- What shapes did you identify?
- What are the characteristics of each shape?

Usage Notes: Emphasize the "characteristics" cell for each shape; students should name qualities of sides, their relationships, area formulas, and so forth.

THGS-3: Polyhedrons

Objective: Students will define the term *polyhedron,* will identify the parts of a polyhedron, and will identify the five regular polyhedrons.

Key Questions:
- What is a polyhedron?
- What are sides? vertices?
- What is a convex polyhedron?
- What are the five regular polyhedrons?

Usage Notes: Extend the activity sheet by having students sketch the five regular polyhedrons (which are also called the *Platonic solids*).

THGS-4: Cubes

Objective: Students will define the term *cube* and identify and discuss the formulas for a cube's surface area and volume.

Key Questions:

- What is a cube?
- How do you determine the surface area of a cube?
- How do you determine the volume of a cube?

Usage Notes: Encourage students to memorize the information in the graphic organizer.

THGS-5: Prisms

Objective: Students will define the term *prism* and will identify the parts, types, and uses of a prism.

Key Questions:

- What is a prism?
- What are the parts of a prism?
- What are the types of prisms?

Usage Notes: Consider accompanying work on this graphic organizer with a brief demonstration of the use of an optical prism.

THGS-6: Pyramids

Objective: Students will define the term *pyramid*, identify its parts, and learn the formulas for the surface area and volume of a pyramid.

Key Questions:

- What is a pyramid?
- What are the parts of a pyramid?
- How do you determine the surface area and volume of a pyramid?

Usage Notes: Explain to students that the four-sided pyramid with which they are most familiar is only one of an infinite number of types of pyramids and is actually a tetrahedron.

THGS-7: Spheres

Objective: Students will define the term *sphere*, identify its parts, and learn the formulas for the surface area and volume of a sphere.

Key Questions:
- What is a sphere?
- What are the parts of a sphere?
- How do you determine the surface area and volume of a sphere?

Usage Notes: Extend the activity sheet by having students discuss its information using a three-dimensional sphere as a visual aid.

THGS-8: Cylinders

Objective: Students will define the term *cylinder,* identify its parts, learn the formulas for the surface area and volume of a cylinder, and identify four types of cylinders.

Key Questions:
- What is a cylinder? What are four types of cylinders?
- What are the parts of a cylinder?
- How do you determine the surface area and volume of a cylinder?

Usage Notes: Reinforce the lessons of the activity sheet by having students transfer its information onto an old coffee can covered with paper.

THGS-9: Cones

Objective: Students will define the term *cone,* identify its parts, and learn the formulas for the surface area and volume of a cone.

Key Questions:
- What is a cone?
- What are the parts of a cone?
- How do you determine the surface area and volume of a cone?

Usage Notes: If appropriate, extend the Taking Another Step activity by providing a brief overview of the concept of conic sections.

THGS-10: Surface Area

Objective: Students will define the term *surface area* and explain how to determine the surface area of various solids.

Key Questions:
- What is surface area?
- How do you determine the surface area of each of the solids?
- Why is surface area so called?

Usage Notes: Have students share their results from the Taking Another Step activity. Emphasize that there are many real-world uses for surface area.

THGS-11: Volume

Objective: Students will define the term *volume* and explain how to determine the volume of various solids.

Key Questions:
- What is volume?
- How do you determine the volume of each of the solids?

Usage Notes: Have students share their results from the Taking Another Step activity. Emphasize that there are many real-world uses for volume.

Name _____ Date _____ | **Student Activity Sheet**

THGS-1: What Is a Solid?

Introduction: In geometry there are two basic kinds of shapes. Two-dimensional shapes, or flat shapes, exist on a plane. **Three-dimensional shapes,** or **solids,** exist in three-dimensional space. Learn more about them by following the directions below.

Directions: Use your textbook, talk to your teacher and classmates, and conduct other research to complete the diagram.

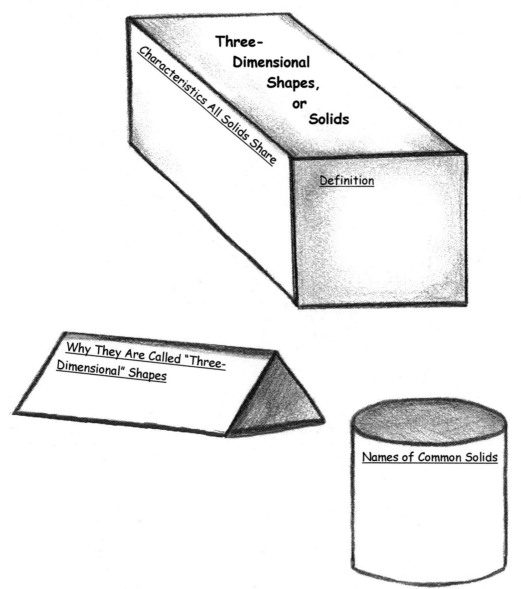

Three-Dimensional Shapes, or Solids

Characteristics All Solids Share

Definition

Why They Are Called "Three-Dimensional" Shapes

Names of Common Solids

Taking Another Step: What are the names of the solids in the diagram? Write your answers next to each one.

Name _____ Date _____

THGS-2: Characteristics of Three-Dimensional Shapes

Introduction: There are many different types of solids, each with its own characteristics. Knowing what makes each type of solid unique is vital to your study of geometry.

Directions: Complete and expand the diagram by adding boxes to the vertical line as you learn about different types of three-dimensional shapes.

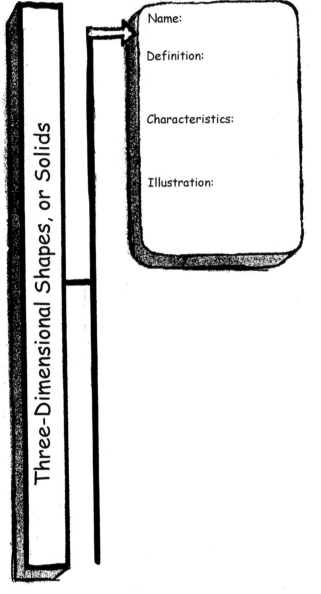

Taking Another Step: On the back of this sheet, write down the names of as many different three-dimensional shapes as you can see in your classroom or in your home.

29   *Graphic Organizers for Geometry*

Name_____ Date _____

THGS-3: Polyhedrons

Introduction: **Polyhedron** is a fancy word, but polyhedrons themselves are easy to understand—as you are about to see.

Directions: Use your textbook, talk to your teacher and classmates, and conduct other research to complete the diagram.

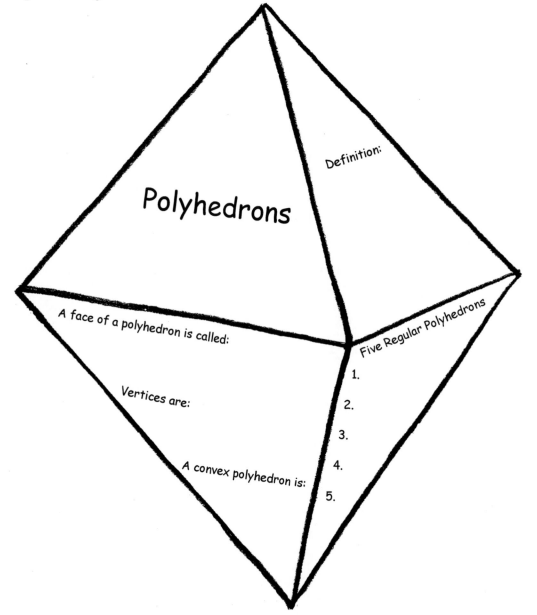

Taking Another Step: What type of polyhedron is represented by the diagram? Is it convex? Write your answers on the back of this sheet.

   Graphic Organizers for Geometry

Name_____ Date _____

THGS-4: Cubes

Introduction: You are already familiar with the kind of regular polyhedrons called **cubes.** Dice are cubes, many boxes are cubes, ice comes in cubes. Cubes are a common shape in architecture. In fact, cubes are all around you. Knowing about them will help you understand your world just a little bit better.

Directions: Use your textbook, talk to your teacher and classmates, and conduct other research to complete the diagram.

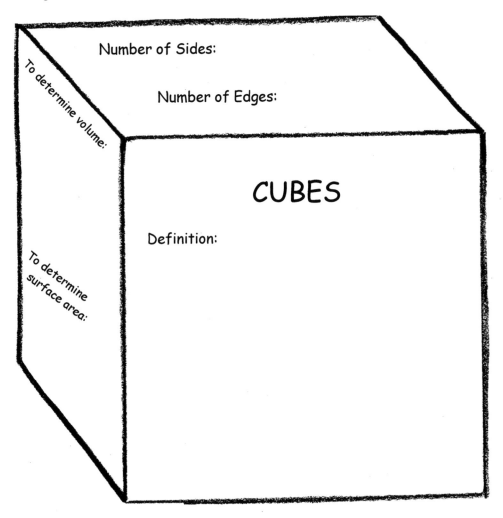

Taking Another Step: On the back of this sheet, write a paragraph explaining how understanding cubes could help one of the following people: an architect, an artist, a scientist, a truck driver.

Name_____ Date _____

THGS-5: Prisms

Introduction: The word **prism** has a strange origin. It comes from an old Greek word meaning "thing sawed off"! Can you guess why? After you've done enough work to complete the diagram below, you might just figure it out!

Directions: Use your textbook, talk to your teacher and classmates, and conduct other research to complete the diagram.

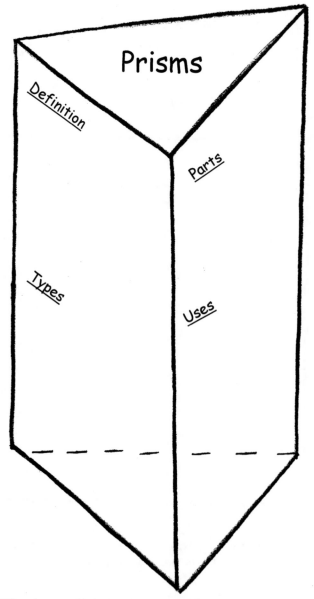

Taking Another Step: Why do you think the Greeks called prisms "things sawed off"? Write your answer on the back of this sheet.

                   Graphic Organizers for Geometry

Name _____ Date _____

THGS-6: Pyramids

Introduction: You probably know about the **pyramid** from the famous pyramids built in ancient Egypt. Pyramids were also built by other peoples, in other places, in other times. It might surprise you to learn that the four-sided pyramid is not the only kind of pyramid—far from it.

Directions: Use your textbook, talk to your teacher and classmates, and conduct other research to complete the diagram.

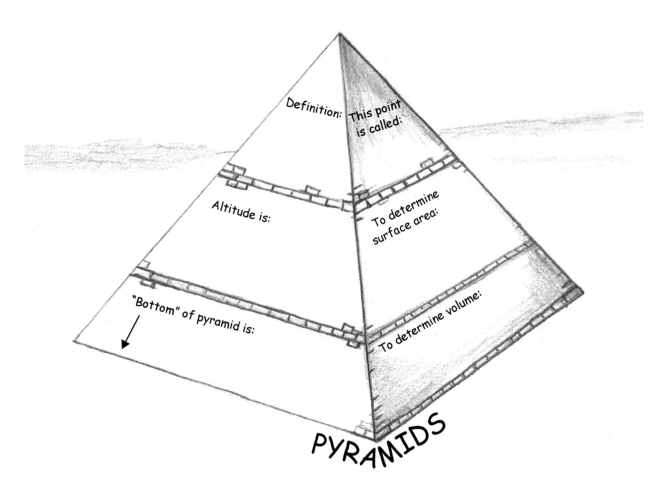

Taking Another Step: The pyramid in the diagram can be called other names. What are they? (*Hint:* Think about terms for solids.)

Name _____ Date _____ | **Student Activity Sheet** |

THGS-7: Spheres

Introduction: There are an infinite number of solids, but only a few that have names. One of them is the **sphere.** Our word sphere comes from the Greek word for "ball." Spheres are also called **balls** and **globes.**

Directions: Use your textbook, talk to your teacher and classmates, and conduct other research to complete the diagram.

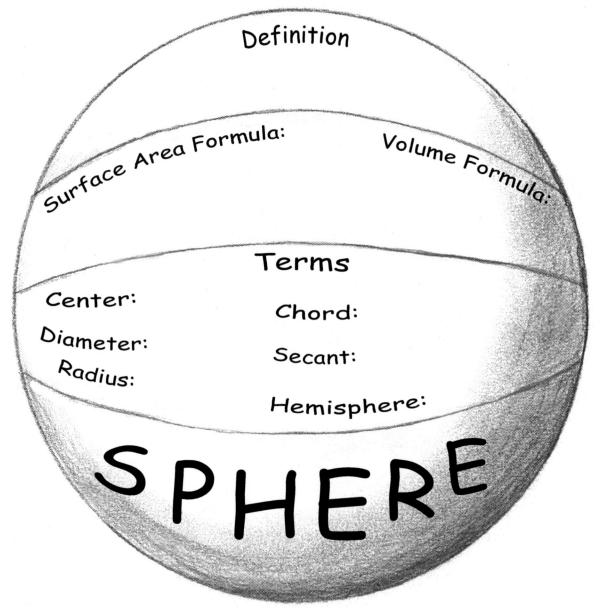

Taking Another Step: On the back of this sheet, list ten examples of spheres.

 Graphic Organizers for Geometry

Name_____ Date _____ |

THGS-8: Cylinders

Introduction: **Cylinders** are common shapes. You see cans and barrels, for example, all the time. But how much do you really know about cylinders? You're about to find out!

Directions: Use your textbook, talk to your teacher and classmates, and conduct other research to complete the diagram.

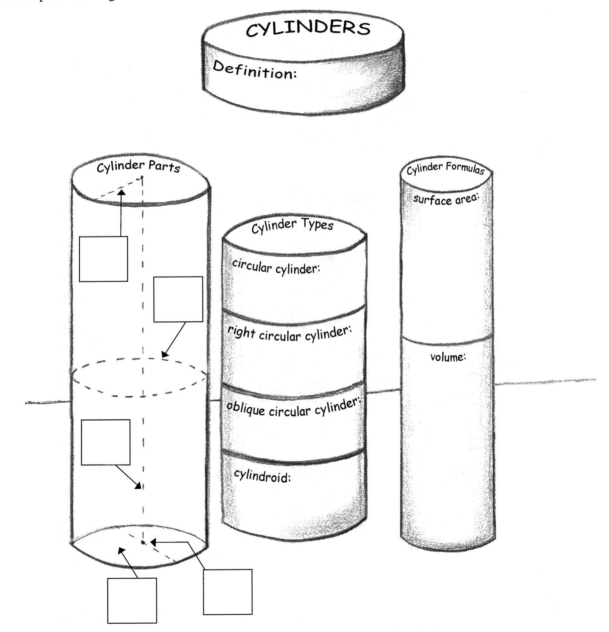

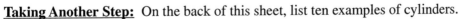

Taking Another Step: On the back of this sheet, list ten examples of cylinders.

Name_____ Date _____ |

THGS-9: Cones

Introduction: If for no other reason, **cones** are important because they hold ice cream! Of course, they're important for other reasons, too. Think about how cones are used as you complete the diagram.

Directions: Use your textbook, talk to your teacher and classmates, and conduct other research to complete the diagram.

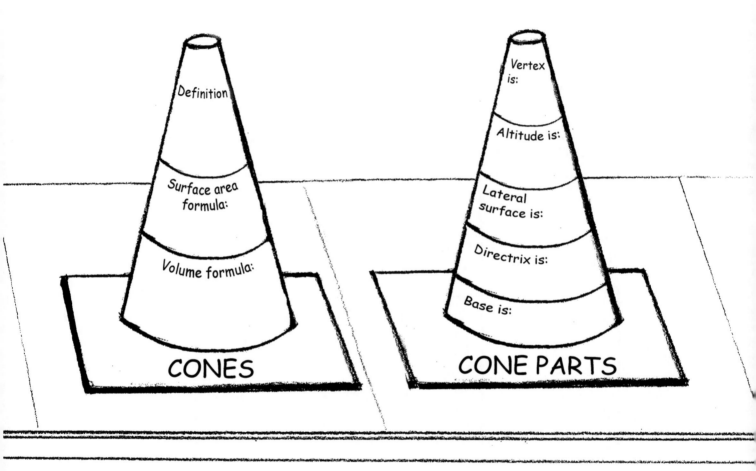

Left cone (CONES): Definition / Surface area formula: / Volume formula:

Right cone (CONE PARTS): Vertex is: / Altitude is: / Lateral surface is: / Directrix is: / Base is:

Taking Another Step: How are circles, ellipses, and parabolas related to cones? Write your answer on the back of this sheet.

     Graphic Organizers for Geometry

Name _____ Date _____

THGS-10: Surface Area

Introduction: **Surface area** is a concept you need to know about to understand solids. A key thing to remember about surface area is that it is expressed in **square measure.** Remember this:

> The unit for surface area is the square.

Directions: Complete the table.

Surface Area	
Definition	
Determining Surface Area	
Solid	*Formula or Process for Determining Surface Area*
polyhedron	
cube	
prism	
pyramid	
sphere	
cylinder	
cone	
other	
other	
other	
other	

Taking Another Step: Think of three real-life situations in which knowing the surface area of something is important. On the back of this sheet, tell about each one.

             Graphic Organizers for Geometry

Name _____ Date _____ | **Student Activity Sheet**

THGS-11: Volume

Introduction: **Volume** is a concept you need to know about to understand solids. A key thing to remember about volume is that it is expressed in **cubic measure.** Remember this:

> The unit for volume is the cube.

Directions: Complete the table.

Volume	
Definition	
Determining Volume	
Solid	**Formula or Process for Determining Volume**
polyhedron	
cube	
prism	
pyramid	
sphere	
cylinder	
cone	
other	
other	
other	
other	

Taking Another Step: Think of three real-life situations in which knowing the volume of something is important. On the back of this sheet, tell about each one.

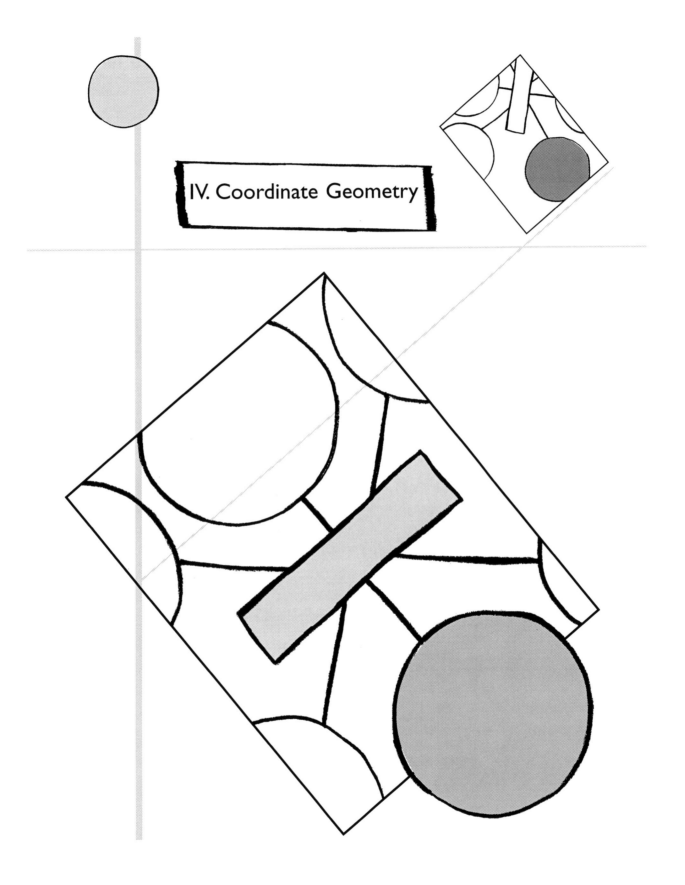

IV. Coordinate Geometry

CG-1: What Is Coordinate Geometry?

Objective: Students will answer six basic questions about coordinate geometry.

Key Questions:

- Questions on the graphic organizer.

Usage Notes: Consider using this graphic organizer as an outline for an introductory lecture on coordinate geometry.

CG-2: The Coordinate Plane

Objective: Students will identify and discuss the elements of the coordinate plane.

Key Questions:

- What is the coordinate plane?
- What are the parts of the coordinate plane?

Usage Notes: Use this activity sheet in conjunction with AGT-5, the Coordinate Plane Template.

Name_____ Date _____

CG-1: What Is Coordinate Geometry?

Introduction: **Coordinate geometry** tells you what it is: geometry using coordinates—places on the coordinate plane. You've advanced far enough in your study of geometry to understand coordinate geometry with just a little effort. This activity will help you.

Directions: Use your textbook, talk to your teacher and classmates, and conduct other research to answer the questions.

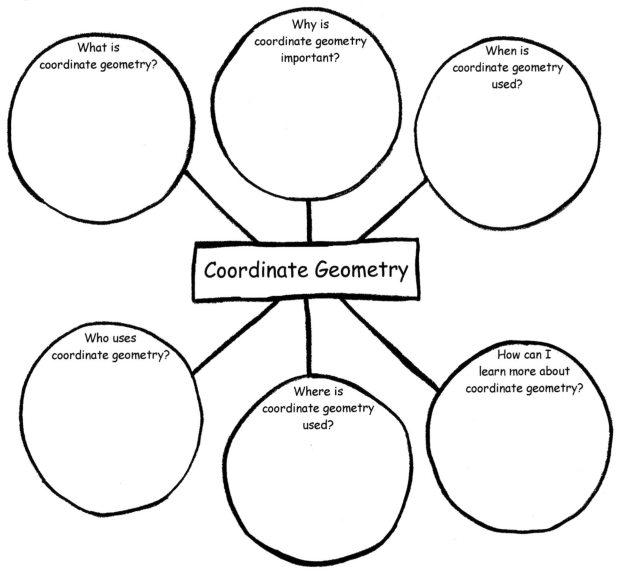

Taking Another Step: On the back of this sheet, write any other questions you have about coordinate geometry. Then work with your teacher and classmates to answer them.

                           Graphic Organizers for Geometry

Name_____ Date _____ | **Student Activity Sheet**

CG-2: The Coordinate Plane

Introduction: Coordinate geometry needs coordinates, of course . . . but coordinates on what? Enter the **coordinate plane.**

Directions: Use your textbook, talk to your teacher and classmates, and conduct other research to complete the diagram.

The Coordinate Plane

Definition: _____

_____

Coordinates

$$(x, y)$$

This coordinate is called:

This coordinate is called:

This number line is called:

This number line is called:

Taking Another Step: On the back of this sheet, write a paragraph explaining how algebra is an important part of coordinate geometry.

 Graphic Organizers for Geometry

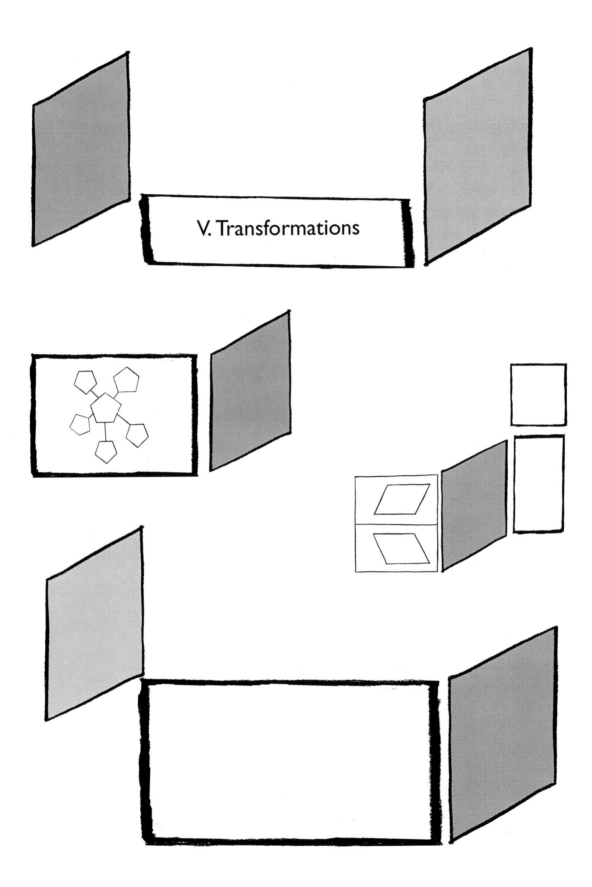

V. Transformations

T-1: Symmetry

Objective: Students will define the term *symmetry* and identify three fundamental concepts of symmetry.

Key Questions:
- What is symmetry?
- What is a plane of symmetry?
- What is an axis of symmetry?
- What is a center of symmetry?

Usage Notes: Emphasize to students that there are different types of symmetry.

T-2: Similarity

Objective: Students will define the term *similarity* and identify fundamental concepts of similarity.

Key Questions:
- What is similarity?
- How is similarity used?
- How is similarity different from congruence?

Usage Notes: Emphasize to students the concept of proportionality as it applies to similarity.

T-3: Congruence

Objective: Students will answer five fundamental questions about congruence.

Key Questions:
- Questions on the graphic organizer.

Usage Notes: Emphasize the importance of congruence in real-world applications (e.g., part fitting in mass production).

T-4: Rotations, Slides, and Flips

Objective: Students will identify and explain basic plane figure transformations.

Key Questions:
- What is a rotation? a slide? a flip?
- Do these transformations result in congruent figures?

Usage Notes: Help students apply this knowledge through the use of tracing paper, mirrors, or geometry software.

Name _____ Date _____ | **Student Activity Sheet**

T-1: Symmetry

Introduction: Like many other terms in geometry, the word **symmetry** has an interesting—and telltale—origin. Symmetry comes from a Greek word meaning "of like measure." Think about why this is so as you complete the diagram.

Directions: Use your textbook, talk to your teacher and classmates, and conduct other research to complete the diagram.

Taking Another Step: Why is the origin of the word *symmetry* appropriate? Write your answer on the back of this sheet.

45   *Graphic Organizers for Geometry*

Name_____ Date _____

T-2: Similarity

Introduction: **Similarity** is an important idea for understanding flat shapes. This exercise will help you. So will simply remembering that similar figures are similar—but not exactly alike.

Directions: Use your textbook, talk to your teacher and classmates, and conduct other research to complete the diagram.

Taking Another Step: All the triangles in the diagram are *similar.* What does this mean? Write your answer on the back of this sheet.

   Graphic Organizers for Geometry

Name_____ Date_____

T-3: Congruence

Introduction: You may not know the word **congruence,** but you've understood what it meant ever since you were a little kid playing with models or dolls. As you complete this activity, you'll understand what that means.

Directions: Use your textbook, talk to your teacher and classmates, and conduct other research to complete the diagram.

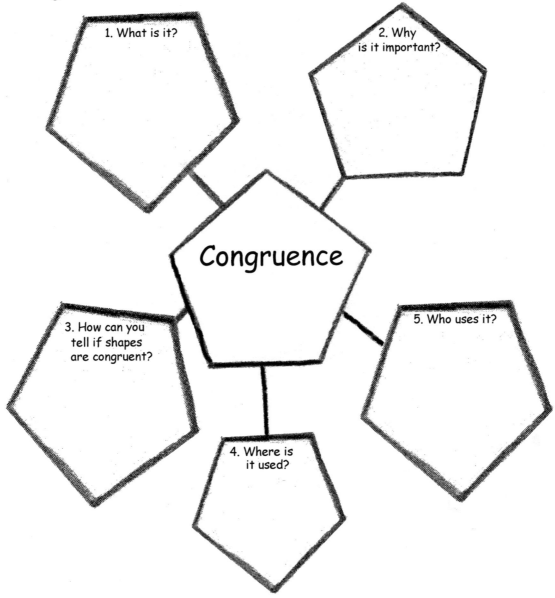

Taking Another Step: Which of the pentagons in the diagram are congruent? How do you know? Write your answer on the back of this sheet.

47   *Graphic Organizers for Geometry*

Name_____ Date _____ | **Student Activity Sheet**

T-4: Rotations, Slides, and Flips

Introduction: Part of the fun of geometry is moving stuff around. With plane figures, you can push, pull, twist, spin, flip—have a ball! Meanwhile, you'll be learning some important ideas. Because these movements change, or **transform,** the figures, they are called **transformations.**

Directions: Use your textbook, talk to your teacher and classmates, and conduct other research to complete the table.

Transformations				
Type of Transformation	**Other Name(s) for It**	**Definition**	**Results in a Congruent Figure?**	**Example (draw a figure before and after the transformation)**
Rotation				
Slide				
Flip				

Taking Another Step: Draw five figures and examples of their transformations on the back of this sheet. Trade sheets with a classmate and challenge each other to identify what types of transformations are illustrated.

VI. Communication and Problem Solving

CPS-1: Problem-Solving Strategies

Objective: Students will identify and analyze four generic problem-solving strategies.

Key Questions:

- What problem-solving strategies did you identify?
- How does each one work?
- When have you used each one?

Usage Notes: Problem-solving strategies to consider include: looking for patterns; solving a similar, simpler problem; breaking the problem into smaller parts; working backwards; making a sketch; and so on.

CPS-2: K-W-L Chart

Objective: Students will apply a K-W-L chart to a mathematics activity or assignment.

Key Questions:

- Questions on the graphic organizer
- How can you use a K-W-L chart in the future?

Usage Notes: Encourage students to see the value of K-W-L charts, and to use K-W-L charts repeatedly in their academic careers.

CPS-3: Problem-Solving Myths

Objective: Students will identify five common myths about problem solving.

Key Questions:

- Why do you think each myth has developed?
- What truth should replace each myth?
- What steps can you take to concentrate on the truth and forget the myth?

Usage Notes: Consider assigning this graphic organizer very early in the school year.

CPS-4: Myself as a Problem-Solver

Objective: Students will identify their strengths and weaknesses as problem-solvers and determine ways to overcome their weaknesses.

Key Questions:

- What are your strengths as a problem-solver?
- What are your weaknesses as a problem-solver?
- What can you do to overcome each weakness?

Usage Notes: Be careful to respect students' privacy. Suggest specific, concrete ways for individual students to address each problem-solving weakness.

CPS-5: Geometric Communication Skills

Objective: Students will identify and discuss the use of communication skills in a geometric context.

Key Questions:
- What are good geometric communication skills?
- Why is each one important?
- How are these skills related to life outside geometry class?

Usage Notes:
- This graphic organizer works well as a whole-class activity.
- Cite, or have students cite, classroom examples of each skill being put to use.

CPS-6: A Letter Explaining a Geometric Concept

Objective: Students will write a letter explaining a geometric concept, then evaluate how well they communicated.

Key Questions:
- How difficult do you think this assignment will be?
- What made it easy or difficult for you?
- How well did you communicate?
- How does expressing a geometric concept in writing help you understand and remember it?
- How does expressing a geometric concept in writing help you understand the use of symbols in geometry?

Usage Notes:
- This graphic organizer works well as a homework activity.
- Assign individual students different concept topics. Consider assigning concepts that particular students are having difficulty with.

CPS-7: Working in a Group

Objective: Students will identify and apply seven group-work skills.

Key Questions:
- What group-work skills did you identify?
- Why is each skill important?
- How can you be sure to apply each skill?
- What should you do if someone else in the group is not using these skills?

Usage Notes: Consider assigning this graphic organizer in conjunction with a specific group project.

Name_____ Date _____ | **Student Activity Sheet**

CPS-1: Problem-Solving Strategies

Introduction: How do you solve a geometry problem when no one has shown you how to do it? The answer is that you apply a problem-solving strategy that you already know. **Problem-solving strategies** are general ways to approach problems to find solutions. Even if you don't know the specific way to solve a problem, a general strategy can put you on the right path and lead you to a solution.

Directions: Think about the different paths you have taken to solve problems in the past. Then complete the diagram by identifying and describing a general problem-solving strategy on each path.

Taking Another Step: On the back of this sheet, write about a time when you applied one of these problem-solving strategies successfully.

Name_____ Date _____ | **Student Activity Sheet** |

CPS-2: K-W-L Chart

Introduction: K-W-L stands for What I **K**now, What I **W**ant to Know, and What I **L**earned. Filling out a K-W-L chart before, during, and after a geometry assignment is a good way to keep track of your progress.

Directions: Complete the K-W-L chart.

Activity or Assignment: _____		
K (complete before)	**W** (complete before)	**L** (complete during and after)
What do you already know about the activity or assignment?	What do you want to learn about the activity or assignment?	What did you learn about the activity or assignment?

Taking Another Step: On the back of this sheet, write a paragraph explaining how using a K-W-L chart did or did not help you with your assignment.

Name _____ Date _____   **Student Activity Sheet**

CPS-3: Problem-Solving Myths

Introduction: Geometry teachers know that, when it comes to solving geometry problems, many students are their own worst enemies. This is because many students hold mistaken beliefs about problem solving. These myths can make geometry more difficult than it needs to be for you. So take some time and replace these myths with the truth. You will be doing yourself a big favor!

Directions: Complete the table.

Problem-Solving Myths		
The myth	**The truth**	**What I can do to forget the myth and concentrate on the truth**
"All geometry problems can be solved quickly and directly, if you know how."		
"If you can't see **right away** how to solve a geometry problem, you won't be able to solve it."		
"There is one and only one· right way to solve any geometry problem."		
"The best way to solve a geometry problem is to dive right in—sitting and thinking about it beforehand is a waste of time."		
"Either you have a gift for geometry or you don't. If geometry doesn't come easily to you, you will never succeed at it."		

Taking Another Step: Share your completed table with a classmate. Compare and contrast your entries in the last column.

Name_____ Date_____

CPS-4: Myself as a Problem-Solver

Introduction: As a geometry student, you are a problem-solver. And as a problem-solver, you have both strengths and weaknesses. By identifying and thinking about them you can improve your performance in geometry.

Directions: Complete the diagram. Talk to your teacher if you need help.

My Strengths

My Weaknesses

Ways to Overcome Each Weakness and Make It a Strength!

Taking Another Step: Arrange to privately discuss with your teacher any problem-solving weaknesses you have.

 Graphic Organizers for Geometry

Name_____ Date_____ | **Student Activity Sheet**

CPS-5: Geometric Communication Skills

Introduction: If you think about it, you'll realize that communication is a big part of geometry. Your teacher communicates to your class, to groups, and to you individually, both through speaking and through grading. You communicate with your teacher by asking questions, giving answers, and turning in your assignments. You communicate with your classmates when you cooperate to solve problems. You even communicate with yourself!

Directions: Think about what makes for good communication in your geometry class. Then complete the diagram by identifying and describing six good communication skills you should put to work in your geometry class.

Good Communication in Geometry Class

Taking Another Step: Review your diagram. Place a star (*) by the communication skills you do well. For the others, write one thing you can do to improve each one.

 Graphic Organizers for Geometry

Name_____ Date_____

CPS-6: A Letter Explaining a Geometric Concept

Introduction: Your geometry teacher has a tough job. He or she has to explain what can be difficult and complicated concepts. How well would you do it? You might never be a geometry teacher, but playing the part is a good way to see how much you've learned, and to help you remember it.

Directions: In the space below, write a letter to your teacher or to someone else explaining an important concept you learned in geometry class.

Taking Another Step: Give your letter to someone else. After he or she has read it, quiz that person on the concept you explained. Then ask yourself: How well did I communicate this concept? What could I have done to communicate more effectively?

         Graphic Organizers for Geometry

Name _____ Date _____

CPS-7: Working in a Group

<u>Introduction:</u> Working with other students is something you'll do throughout your school career, and beyond. Do you know what it takes to be an effective group member?

<u>Directions:</u> Each student below is having a different thought about what it takes to be good member of a work group in geometry class. Write the things each one is thinking about.

Good Group Skills

<u>Taking Another Step:</u> The next time you work in a group, do your best to apply the skills you identified in the diagram. After your work with the group is finished, write a journal entry in which you evaluate how well you performed as a member of the group. Note anything you can do to improve your performance next time.

                  *Graphic Organizers for Geometry*

VII. Connections

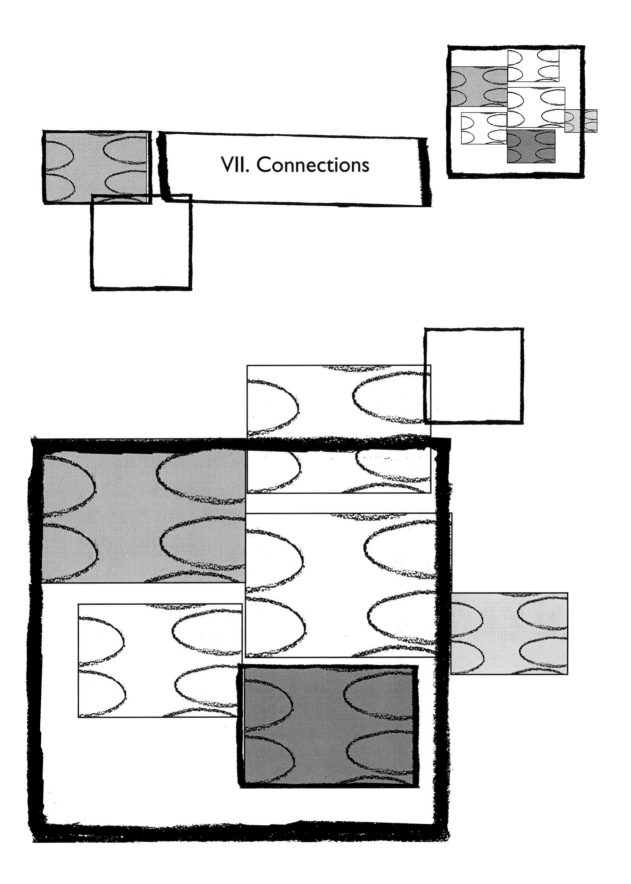

C-1: The Branches of Geometry

Objective: Students will identify and describe the major branches of geometry.

Key Questions:
- What are the main branches of geometry?
- What is each branch concerned with?
- How are these branches similar? How are they different?

Usage Notes:
- This graphic organizer works well as a whole-class activity.
- Consider using this graphic organizer early in the year, as an overview, and having students refer to it periodically to keep them oriented in a macro sense.

C-2: A Geometry Problems Constellation

Objective: Students will create a geometry problems constellation to see the relationships among different geometry problems.

Key Questions:
- How are the problems connected to each other?

Usage Notes:
- This graphic organizer works well as a whole-class activity.
- Be sure to model a geometry problems constellation on the board to ensure student understanding of the concept.

C-3: How I Use Geometry in Other Subjects

Objective: Students will identify ways that geometry is used in other school subjects.

Key Questions:
- In what other subjects do you use, or could you use, geometry?
- Why is geometry important in each subject?

Usage Notes: Consider having students keep geometry-use logs in their other subjects prior to completing this graphic organizer.

C-4: The Golden Rectangle in Art and Architecture

Objective: Students will identify the golden rectangle and investigate its use in art and architecture.

Key Questions:
- What is the golden section? the golden rectangle?
- How are the two related?

Usage Notes: Hold a class discussion on the validity of the concept of the golden rectangle as especially pleasing to the eye.

Name_____ Date _____

C-1: The Branches of Geometry

Introduction: Geometry is divided into many different branches. The different branches of geometry are concerned with different types of problems and how the solutions to those problems are put to use. But the branches, at their root, have more similarities than differences: they're all part of geometry.

Directions: Conduct research to complete the diagram. Make sure you write a description of each branch of geometry and an explanation of its importance.

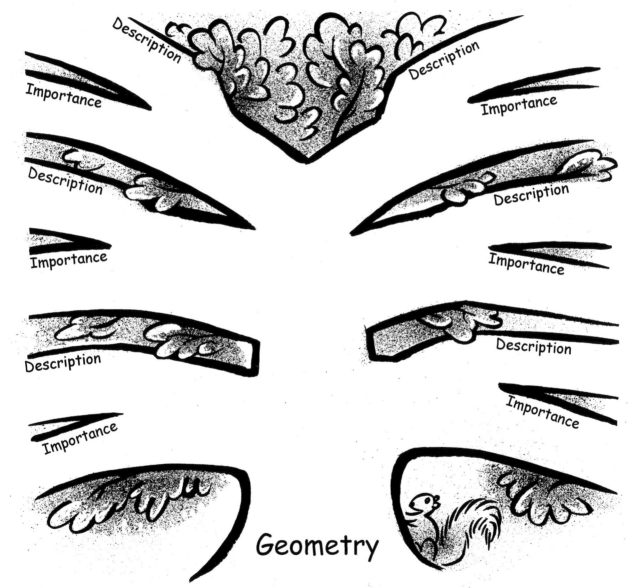

Taking Another Step: Think of an example of a real-world use for each branch of geometry you wrote about above. Write the example on the appropriate branch.

         Graphic Organizers for Geometry

Name_____ Date _____ | **Student Activity Sheet**

C-2: A Geometry Problems Constellation

Introduction: A geometry problems constellation is a diagram that shows the relationships among different geometry problems. Each problem is represented by a circle. Lines and arrows between the circles represent the relationships between the problems. Labels on the lines define the relationships between the problems. You can expand problems constellations to include many, even hundreds, of problems. Just be sure to draw the arrows and label them with the correct relationships.

Directions: In the space below, draw a geometry problems constellation. Include at least three problems.

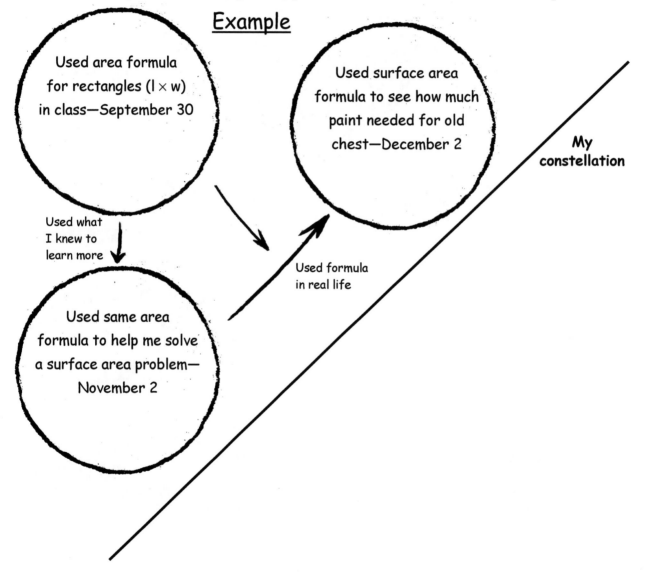

Taking Another Step: Add one problem to your constellation that seems to have absolutely nothing to do with the others. Then challenge yourself to find the connections.

   Graphic Organizers for Geometry

Name_____ Date_____

C-3: How I Use Geometry in Other Subjects

Introduction: Geometry isn't just for geometry class. You will use it in other classes, as well. In fact, you use it just getting *to* class!

Directions: In each classroom on the diagram, write at least two ways you use geometry in that subject.

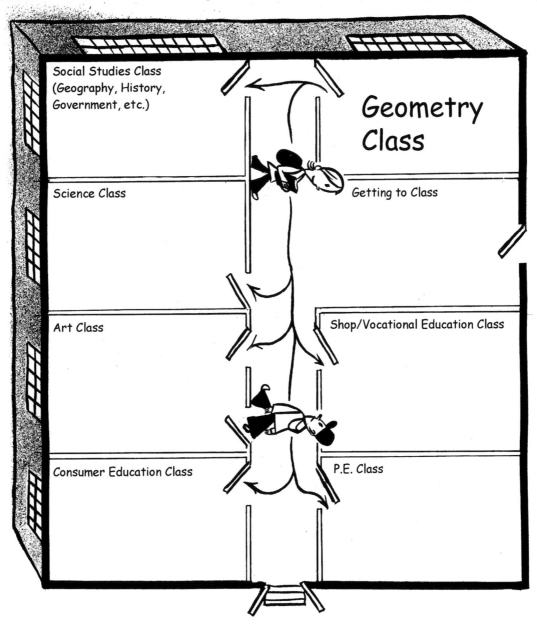

Taking Another Step: On the back of this sheet, tell about one time you used your knowledge of geometry to help you in another subject.

   Graphic Organizers for Geometry

Name_____ Date _____

C-4: The Golden Rectangle in Art and Architecture

Introduction: Look at the rectangle below. Many people believe it is one of the most special shapes known to people. The rectangle before you is a golden rectangle.

A **golden rectangle** is a rectangle with a certain ratio between its length and width. In a golden rectangle, the length is about 1.6 times the width. This ratio is based on the golden section. The **golden section** is the division of a line segment so that the ratio of the whole segment to the larger part is the same as the ratio of the larger part to the smaller part.

What makes the golden rectangle so important? The answer may surprise you. For some reason, the golden rectangle is more pleasing to the eye than other rectangles. It seems more balanced, more beautiful. It just seems right. No one knows why this is so.

Directions: Look for examples of golden rectangles in paintings, architecture, and sculpture. Record any examples you find.

Taking Another Step: Do you find the golden rectangle especially appealing to the eye? Why or why not? Write your answer on the back of this sheet.

64   *Graphic Organizers for Geometry*

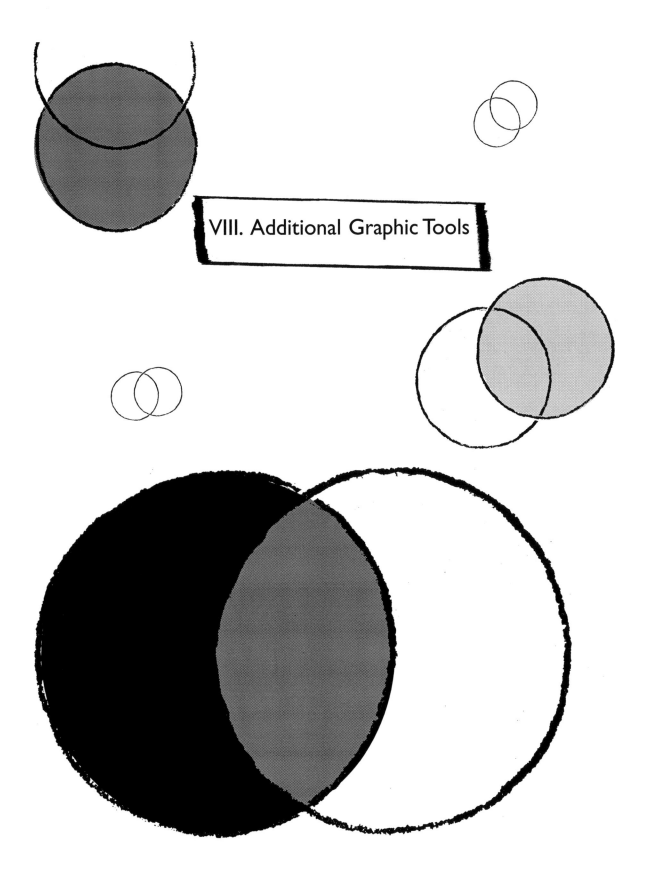

VIII. Additional Graphic Tools

The graphic organizers in this section are self-explanatory. They are also very flexible; you can assign them in a wide variety of classroom situations as you see fit.

Name_____ Date_____

AGT-1: Geometry Symbol Log

Symbol	Meaning

Name_____ Date _____

AGT-2: Geometry Formula Log

Formula	Explanation	Application

 Graphic Organizers for Geometry

Name_____ Date_____

AGT-3: Personal Geometry Glossary

Introduction: A **glossary** is a list of specialized terms and their meanings. As you progress through the school year, you will learn many new terms. This sheet gives you a place to create a personalized glossary of these terms.

Directions: As you learn new terms, record them and their meanings in the table below.

Terms Related to My Study of Geometry	
Term	Definition

Name_____ Date _____

AGT-4: Geometry Use Log

Introduction: You will use geometry in other subjects and in other areas of your life.

Directions: Keep track of how you use geometry outside geometry class.

Using Geometry in Everyday Life	
Date	Description of How I Used Geometry Outside Geometry Class

 Graphic Organizers for Geometry

Name_____ Date _____ |

AGT-5: Coordinate Plane Template

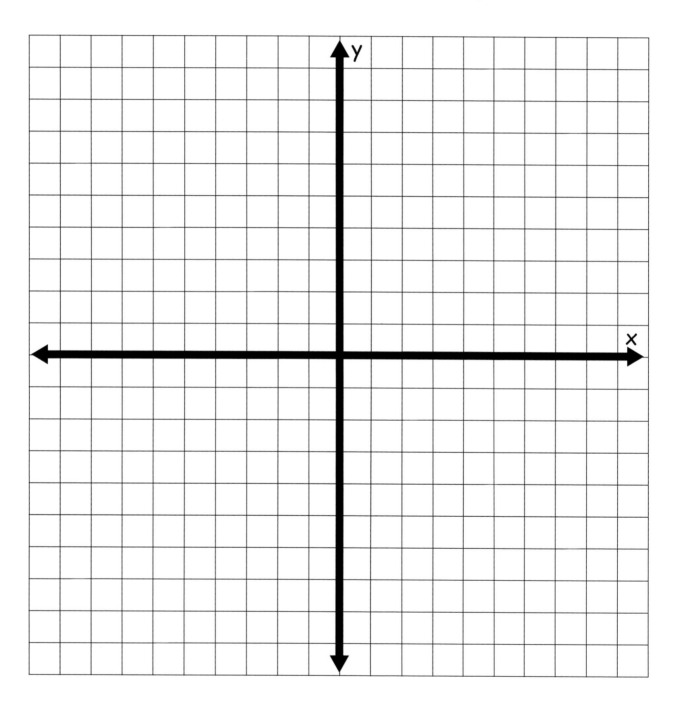

             Graphic Organizers for Geometry

Name _____ Date _____

AGT-6: The Six Questions

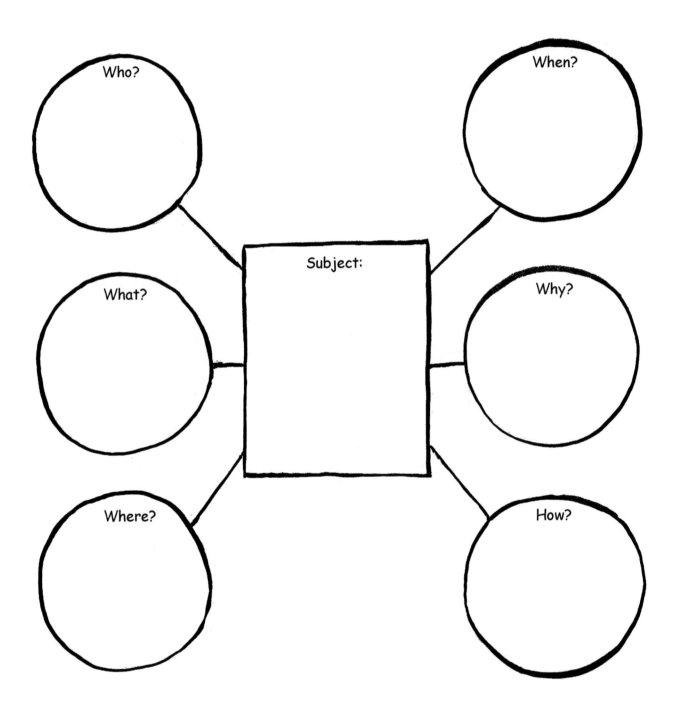

Name_____ Date_____

AGT-7: Venn Diagram Template

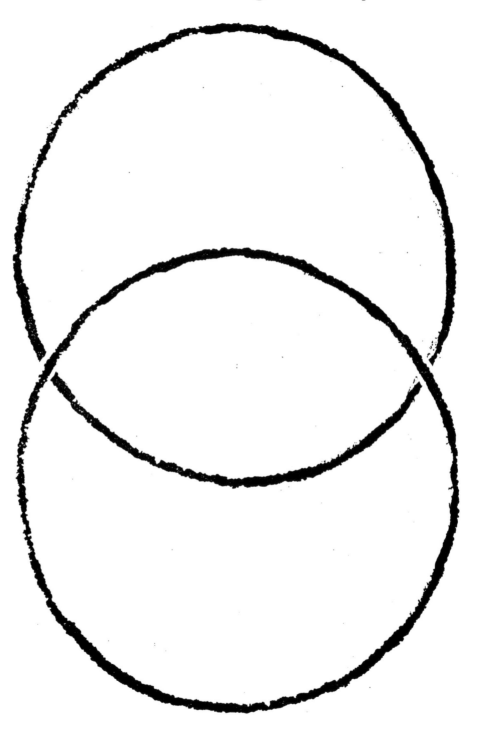

Name_____ Date _____

AGT-8: 10 x 10 Grid

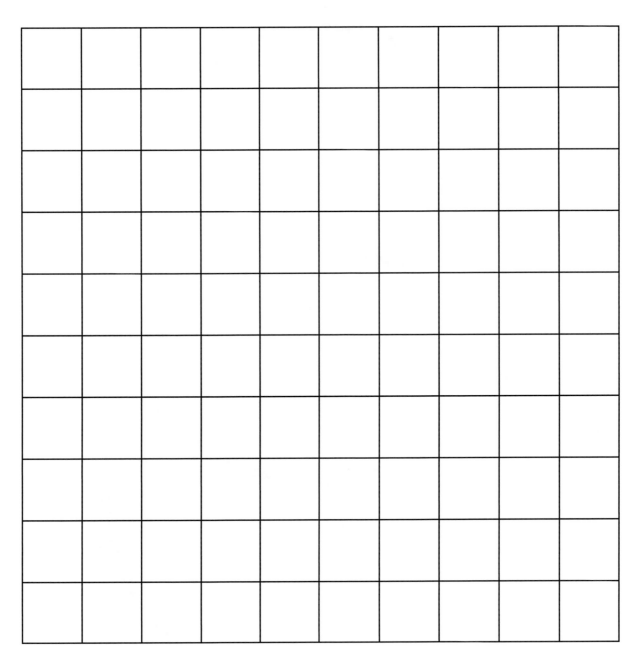

                   Graphic Organizers for Geometry

Name_____ Date_____

AGT-9: 100 x 100 Grid

Graphic Organizers for Geometry

Name_____ Date _____

AGT-10: Dealing with Math Anxiety

Introduction: You're not alone. Many, many students have to deal with **math anxiety.** This term refers to the feelings of fear and frustration you sometimes get when doing math. By identifying these symptoms, and taking steps to eliminate them, you can lower your anxiety a great deal—and even get rid of it entirely.

Directions: List your symptoms on the left-hand side of the diagram. Then work with your teacher to come up with ways to combat them.

Symptoms of My Math Anxiety

Positive Steps I Can Take
to Lessen My Anxiety

76   *Graphic Organizers for Geometry*